The Story of Andre

6.95

The Story of Andre

By Lew Dietz

Illustrated by Stell Shevis

Down East Books / Camden, Maine

Contents

Toni's Note

After Lew Dietz and my father, Harry Goodridge, wrote the book, *A Seal Called Andre,* children from all over the country wrote saying there should be a book about Andre that *they* could read and understand. Lew Dietz, our neighbor, has been a friend of the family since before I was born. One day I said to him, "Now that Andre is famous, I think you should write a book about what it was like for a little girl to grow up with a famous seal." "Toni," he said, "I think you are right." So this is the story I told him about growing up with a seal called Andre.

The Story of Andre

1

The Goodridge Family

My name is Toni. I was seven years old that day in May when a baby harbor seal came to live with the Goodridge family. Andre grew up and became as famous as Smokey the Bear. But Andre is real and his story is a true one.

My father Harry is a tree surgeon, but he's also a scuba diver, who used to spend many hours swimming underwater. Diving deep down in the dark, cold waters of Maine is very lonely and often dangerous. For years he'd been thinking that a harbor seal would make an ideal diving companion.

I thought this was a wonderful idea, as did my brother and sisters. There are five of us. The oldest are the twins, Steve and Sue; then came Carol and Paula, and finally me. Being the baby of the family wasn't always easy. My brother and sisters were usually too busy with their own affairs to find time for me. I guess that was the reason I learned to talk to birds. I found birds understanding and very sensible.

My mother Thalice was understanding and sensible, too, but she wasn't at all sure that adding a baby seal to the family was a good idea, for we already had a houseful of pets. Finally she decided that having one more wouldn't make all that much difference so long as she didn't have to clean up after it.

We had the usual frogs, snakes, and pet mice. Even ants. Steve, who was planning to be an engineer, liked to capture dragonflies and load them down with ants, just to see how many ants a dragonfly could fly off with. When Paula was little, she would put diapers on chickens and push them around in her baby carriage. If the neighbors thought we were all a bit odd, we thought it perfectly natural to have birds and animals for playmates.

I had a pet rooster named Jake, and a fat, fussy hen named Mrs. Cluck-Cluck. When they had been going together for years, I thought it was time they got married. I dressed

Mrs. Cluck-Cluck in a white wedding gown with a hole cut out in the train for her tail feathers. The wedding was held down by the duck pond. Our pigeon, named Walter, was best man. Walter spent all his time around the pond with the ducks. Maybe he thought he was a duck, too.

Then there was a flying squirrel, Charlie, who lived somewhere in our house. We never did find out where. He would come gliding out each night at dusk and join us for supper. He loved bananas.

Reuben was our cock robin. He fell out of his nest one day and Steve brought him home. Reuben could never get enough to eat. We would spend hours collecting and feeding him worms until it seemed he would burst. But he still wanted more worms. He loved splashing water, and whenever I took a bath, he would fly in and sit on the edge of the tub. And Mother couldn't wash dishes without Reuben getting in her way. One night when he fell into the soapy water, he looked a sorry sight.

There was also a one-legged sea gull my father called Sam Segal. A friend, finding him wounded at the dump, had brought him to us to care for. Everyone brought us wounded wild animals and birds. Like Mother, they didn't think one more would make that much difference.

We had a dog, of course, a fat, lazy beagle named Toot, and a pet rabbit named Bugs. Toot didn't know that beagles were supposed to chase rabbits. Bugs would point this out to Toot. When Bugs got bored, which was often, he would nudge Toot out of a snooze on the lawn. Toot would finally get up and give Bugs a chase around the house. Then he would go back to sleep again.

Toot was a great help with Beulah, my pet guinea pig. Beulah wasn't very bright. She'd wander off and get lost. When I'd find her gone, I'd call Toot and say, "Go find Beulah." Obligingly, Toot would track her down with his nose and send her scrambling back to the safety of her pen.

And I had a horse named Tiffany. On fine summer days we'd go off together into the woods and fields, and look for the greenest grass for Tiffany to crop. But mostly I enjoyed luring birds into a little wire pen at the back of the house. I'd scatter birdseed and dried corn along a path and into the pen, and all sorts of birds ate their way into captivity. Of course, I always freed them, for wild birds belong in the air or nesting in trees.

One day Daddy said teasingly, "Those birds you trap can't be very smart. I'd just like to see you catch a really smart bird like a rudder-tailed grackle."

I didn't know what a rudder-tailed grackle looked like, but later that day I saw a large black bird with a long tail pecking his way into my wire pen. That afternoon I showed him to Daddy. "Is that the bird you mean?" After that, he never teased me about my bird collection.

But this is a story about a seal. For me, the story begins that day in May when Andre arrived from the sea. And I'm sure that was when the story began for Andre, too.

2

The New Arrival

A seal is not a fish. It is a fish-eating mammal that breathes air, and mother seals have glands that provide milk to feed their babies.

Seals belong to a group of mammals that enter the sea. Whales and dolphins, for instance, live entirely in the water, but seals and sea otters bear their young on the land and enter the sea in search of food.

Five kinds of seal can be seen in Maine waters, but the harbor seal is the only year-round resident. Having had their babies, harbor seals spend most of their days on weed-covered tidal ledges in the bays of Maine. Seal mothers have these babies, called pups, in the spring, around the middle of May.

It was on a Saturday in mid-May that my father went out into the bay in his boat to find a baby seal. We were all home and waiting when we heard his truck come into the driveway. We rushed to the window. Sue yelled, "He's done it! He's got one!"

There was Daddy coming up the walk, and peering out from under his slicker was a tiny whiskered head no bigger than a softball. Marching into the kitchen and placing the little creature at our feet, he showed us the seal's glossy black coat, smooth as a grape. I'd never seen anything so beautiful.

When we all began chattering at once, Daddy said, "Now give the little fellow a chance to get used to you funny-looking people." But the little seal seemed not at all disturbed by the fuss. After looking us over one by one, he went bellying about the room examining every nook and cranny, curious as a kitten.

While Andre explored his new home, my father told us what had happened. That morning the sea had been very calm. In his boat he cruised five miles into Penobscot Bay from Rockport, where we live. Eider ducks flashed by; a gull soared into the sky. As my father steered around a weedy ledge, he suddenly saw a tiny head pop up. It was a pup

seal, no more than three days old. Easing the boat closer, my father expected the pup to submerge, but instead the little head came up out of the water as if to get a better look at the strange man in a boat. The baby seal's mother was nowhere to be seen, a surprising thing, for seal mothers are very protective of their young.

Then a curious thing happened. Instead of slipping under water, the little creature began swimming directly to the boat, and it was then that Daddy simply took out his net and scooped the baby seal into the boat. "That's what made me think he was an orphan," my father explained. "Seal mothers usually have only one pup. Occasionally, a mother has two. But since she doesn't have enough milk to feed two babies, one must be abandoned to starve to death or to be eaten by a shark." Hearing that made me glad that my father had come along at the right moment.

I didn't want to think about what had happened to our other seals. You see, we had had two others, first Marky, then Basil. I like stories with happy endings, but the stories of those two both had unhappy endings. Still, it's important to tell what happened to them. I remember overhearing my mother say one day to a neighbor, "Bringing up the first two children was the hardest. After that it got easier, for you learn from your mistakes."

My father never would have been able to bring up the seal we called Andre if he hadn't learned from the mistakes he made with Marky and Basil.

3

Marky and Basil

Mark was named for an island in Penobscot Bay, where he was born. My father knew, of course, that a seal must have its mother's milk, which is ten times richer than cow's milk.

So to canned cow's milk he added egg yokes, and tried in every way to get little Marky to drink this formula. Human mothers can read books on the care and feeding of babies, but there was no book on the care and feeding of baby seals.

As Marky got thinner and thinner, Daddy remembered the pet squirrels he had raised when he was a small boy. His three baby squirrels, Tom, Dick, and Harry, wouldn't drink their milk, either. Then, he thought, "Squirrels like to gnaw on things." So he soaked a dishrag in milk and, sure enough, the baby squirrels began chewing on the dishrag, and grew fat and healthy. So now he soaked a dishrag in milk for Marky. Marky chewed on the rag and began to perk up.

But not for long. One morning we found Marky dead. Later, our animal doctor friend, Mac, performed an operation on the baby seal, and found a dishrag in little Marky's stomach.

My father decided to try once more to bring up a baby seal for a diving companion. But first he set out to learn all he could about harbor seals. One fall day while cruising in the bay, he came upon a mother seal that had been killed by rifle fire. After examining her body very carefully, he discovered why he had failed to get little Marky to drink his milk.

Seals move about on their bellies. The seal mother's nipples, where their babies drink, are set into shallow little cups and protected by a doughnut-shaped fatty tissue so that the nipples won't be damaged as she slides around on rocks. Seal babies must push their noses into those little cups to get milk.

The next winter my father set about making a seal-feeding station. He first took a short log and hollowed out a hole in the middle of it; then he covered the log with the rubbery material that scuba divers' suits are made of. The hole left a dimple in the log. Under the dimple he put a syringe filled with the milk formula. He made a small hole in the rubbery dimple for the tip of the syringe.

I was doing my homework in the living room the night he finished the feeding station. He called me into the kitchen. "Toni," he asked, "you want to try being a baby seal?"

"Me?"

"Yes, you. You're the closest in size to being a baby seal."

Thinking that being the smallest in the family certainly had its drawbacks, I got down on my hands and knees and pushed my mouth into the rubbery dimple. I sucked on the tip of the syringe.

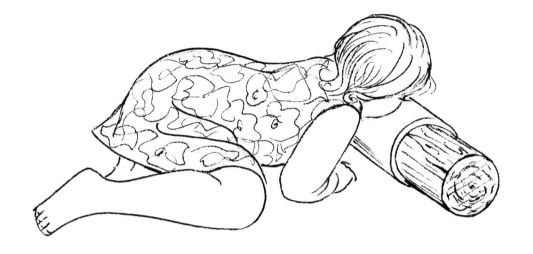

Startled, I sat up. "Great, it works! But do you think it can fool a baby seal?"

"First," Daddy said, "we have to find a baby seal."

Basil was found on the fourteenth of May. Daddy took my sister Sue with him that day. I wanted to go, but was told I wasn't big enough to be much help.

"I tell you what," my father said, "if we get a baby seal, you can be the one to show him how to drink his milk."

I didn't think that was very funny.

They found Basil off Mouse Island. Sue brought him into the house, cuddled in her arms. Sue was sixteen, and had a way with little animals even as I had way with birds. While she talked soothingly to little Basil, Daddy quickly made up a formula of milk and eggs, and filled the syringe.

We all stood around anxiously as he set up the feeding station. This would be the great test. If Basil refused to drink his milk, my father would never again try to bring up a baby seal. Two failures would be enough.

And for a moment it looked very much as if we would never have a baby seal to raise.

Sue set Basil down beside the rubber-covered log. The little seal nuzzled it a moment, then backed away.

"Hah!" Daddy said, "seal mothers are warm. I'll fix that!" So he heated the dimple with warm water from the tea kettle.

That did it! Cautiously, the little seal pushed his nose into the dimple. When he found the warmth, he pushed harder and found the nipple. In an instant he was taking milk like a hungry puppy.

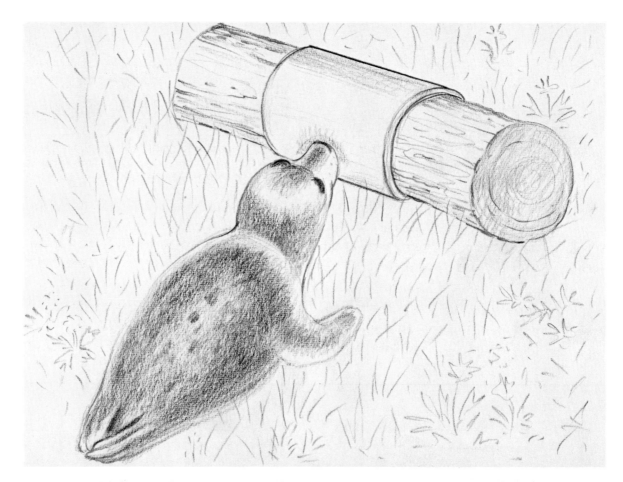

Sue yelled. We all yelled, and began cheering like kids at a basketball game. We made so much noise that Toot, the lazy beagle, woke from his nap and came in to see what the fuss was about. He took one look at the little seal and went back to his snooze. It was almost impossible to impress Toot.

Little Basil grew and grew. If our first problem had been to get him to drink milk, our next was to get him to *stop* drinking it. When the time came for him to be weaned, that is, to give up milk and start eating fish, my father ground some up and added it gradually to the seal's formula. But each time Basil tasted fish, he spit it out in disgust.

By the Fourth of July, Daddy decided that, like it or not, little Basil was going to eat fish, the only natural food for seals. My uncle was with us for our dinner of fresh salmon, green peas, and strawberry shortcake, always eaten in Maine on the Fourth. After dinner, Daddy said to my uncle, "Basil's going to have raw herring for his Fourth of July meal. I may need your help." So while my uncle held little Basil, my father forced the seal's mouth open, pushed a herring into it, and clamped the mouth shut until Basil was forced to swallow.

Basil snorted in disgust. Another fish was forced into his mouth. This time Basil blinked his eyes happily as if to say, "Why, that's not so bad after all." Then he started following Daddy around the house begging for more fish, and thereafter never even looked at his milk-feeding station. It was then that Basil stopped being a baby.

Basil became a wonderful companion. Each day my father would put him in the back seat of the car and take him down to the sea. They would swim and dive together in the harbor, Basil never wandering very far. Underwater, he would come and push his face close to my father's faceplate and peer in at him. On fine days he would be taken out into the bay to frolic in the shallow waters around the island where he was born.

It was no problem getting Basil back into the boat. Daddy had only to start the motor and the seal would make a dash for the boat, and get lifted into the bow. He didn't want to be left alone in the great ocean.

But now comes the sad part of the Basil story.

One day in early August as he was heading for home, my father looked in the bow and Basil was gone. He stopped the motor, thinking that the little fellow had gone for a quick swim in the sea. Then to his horror he saw blood on the water; next he saw a shark's fin cut the surface of the sea. In an instant, he grabbed his harpoon, and drove the barb into the shark's back. The water boiled as the great fish thrashed to free himself from the barb. Then the shark submerged, and my father found himself being towed out to sea by the angry, wounded shark.

Finally, the harpoon line went slack as the great white shark rolled to the surface

and was still. Sadly, Daddy towed the great fish home to the landing, meeting there a group of children who had come down to the waterfront to see little Basil.

Later Daddy opened up the shark, and there inside the stomach was Basil, in three pieces.

It was a sad time for our family. After what had happened to Marky and Basil, we were all sure Daddy would never again try to bring up a baby harbor seal. To our surprise, he decided to make one last effort.

If Marky and Basil had been adorable little creatures, little orphan Andre was to be special, special from the very beginning.

4

Baby Andre

It didn't take Andre more than a day or two to become a full-fledged member of our family. He drank his milk, and grew fat as a butterball. At night he lived in the cellar in an old bathtub. During the day he had the run of the house and yard.

Most of our other pets got along very well with him. Only Sam Segal, the sea gull, had some doubts about Andre. But sea gulls have bad dispositions, complain a lot, and think only of their stomachs. As for little Andre, he thought everyone was his friend.

Daddy was at his desk one evening when my sister Paula called to him from the yard. "Andre just kicked Toot out of his doghouse!" Sure enough, there was Toot dozing on the lawn, and Andre curled up on the hay in Toot's doghouse.

At least twice each day Daddy would take Andre for a swim. Using his flippers, Andre would hump right into the back seat of the car, and off they would go. It didn't seem at all odd to Andre to live in a house with people and yet be chauffeured to the seaside in the back seat of a car.

Sometimes I would go down to the waterfront with Andre and Daddy. Now you'd think baby seals would love to play and swim in the water, but when Daddy would toss Andre into the sea, he would pop right out again and lie in the sun.

At first, Daddy was puzzled. Then he remembered watching mother seals out on the ledges trying to coax their pups into the water. They would nuzzle their babies towards the sea and, using their flippers, sometimes splash them with water. Finally, if that didn't work, they would push the pups into the water. Maine water is icy cold, and Daddy realized that baby seals don't have thick layers of fat to keep them warm. In good time, he said, little Andre would be ready and eager for a swim.

Within a few weeks Andre said goodbye to his bottle and began eating raw fish. That day didn't come soon enough for Daddy, who had to get up in the middle of the night to make the formula. This amused Mother. For years she was the one who got up at night to warm bottles for five children while my father lay in bed.

Once Andre began to eat fish, life was much simpler for all of us. Sometimes Daddy would take him to the harbor and leave him overnight. I'd go down early in the

morning, and there Andre would be under the town float. Happy to see me, he'd bounce up and down in the water, making soft little pleading sounds.

When I said I thought it was cruel to leave him all alone overnight, my father looked me right in the eye and asked me what I planned to do when I grew up.

I thought a moment. Steve was going to be an engineer; Sue, an animal doctor. Carol wanted to be an actress and travel all over the world on a yacht. Though I most wanted to own a zoo, I just said, "Maybe I'll just grow up."

"Andre will be grown up when he's two or three," said Daddy, "but first he must learn to catch fish for himself and be on his own."

"You mean," I gasped, "we can't keep Andre forever!"

"When Andre is ready to join other seals in the bay, don't you think it would be wrong to keep him?" Daddy asked.

That night in bed I cried, thinking about Andre ever leaving us. It didn't make it any easier knowing that my father was right.

One day in late June when Andre had been left in the harbor overnight, I went down to the waterfront with Daddy. He kept frozen herring in a freezer, and each day would throw a few fish to Andre. As usual, Andre was waiting there under the town float. But what happened was very unusual. When his little whiskered face popped up out of the water, he looked at the herring and said, "No fish today, thank you."

Of course, he didn't say this in words, for animals don't talk. But smart Andre was saying, without words, "Keep your silly herring. I've found fresh fish."

Daddy was terribly excited. "Good boy!" he cried. "Andre, you've learned to catch your own fish!"

The little fellow certainly looked pleased with himself. I tried to look pleased, too, but all I could think was that Andre was no longer a baby. One day soon he would head out to sea and we would never see him again.

5

Andre's First Summer

Though by early July Andre was not an easy armful to carry, Daddy brought him home every other day to be with the family, and to teach him a few simple tricks, like rolling over and clapping with his flippers.

While I hoped that Andre wouldn't ever run off to sea, all the same I was very uneasy. Our cock robin, Reuben, had gone off with a flock of robins one fall; our flying squirrel, Charlie, had simply disappeared into the woods one spring.

Then just when I began to think that Andre was happy being in the Goodridge family, he began running off, though not far. When I was little and told to stay in the yard, I would wander off a short distance. Like little boys and girls, Andre, too, was curious about the world.

At first he would be gone overnight; then he would disappear for a day or two. We worried. It would be natural for him to want to live with other seals, but what if Andre

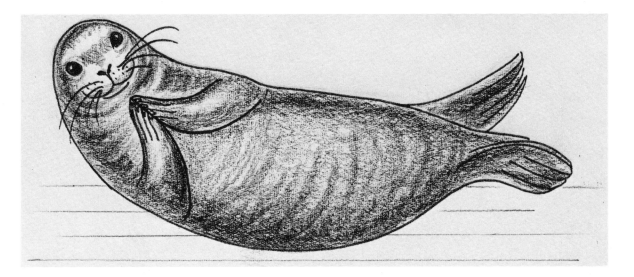

didn't know he was a seal? Wouldn't a little old lady be frightened out of her wits were she to find a seal asleep on her porch?

Once after Andre had been gone three days, my father drove his car down along the coast looking for him. While he was gone, Mother got a phone call from someone in Camden, a town two miles away. Andre was there on the wharf showing off his few tricks to a crowd of people. Mother went to Camden, picked him up, and brought him home.

A few days later he went off again, this time for a whole week. Then we got a phone call from a man in another harbor town. Luckily, he had heard about Andre, and so wasn't startled when he found our seal curled up in his rowboat.

When my father went to pick up Andre, the seal wanted to play. Every time Daddy tried to grab him, Andre dived under the water and popped up again just out of reach.

Annoyed, Daddy said sternly, "Enough of this nonsense, Andre! If you don't want to ride home, you swim home!" So Father drove back to Rockport and waited at the waterfront.

Andre appeared in about half an hour. He must have realized that Daddy had had

23

enough of his nonsense, and made no fuss when he was picked up and taken home.

But that was just the beginning of Andre's friskiness. When he was in the cellar in his bathtub, his favorite trick was to lie low until one of us went by; then he would splash us with his flippers. Once he soaked me when I was dressed in my very best clothes.

I decided it was high time to give the little rascal a dose of his own medicine. A few days later, I went to the waterfront to visit him, carrying a syringe filled with water. When he popped out of the water to say hello, I squirted him squarely in the eyes. He ducked under and came up again. I squirted him once more. If I thought this would teach Andre some manners, I was mistaken. He loved being squirted, and thought it great sport.

Then he decided to play hide-and-seek. One summer day when the tide was low, my sister Sue and I found him in the shallows of Goose River, a little stream that flows into Rockport Harbor. Each time Sue went to grab him, he would swim off and hide behind some floating seaweed. He'd wait for Sue to find him, and then he'd swim off to another hiding place.

Sue, wading out into the water in search of him, happened to look behind her. "Look at him!" she laughed. "The little brat thinks he's hiding." And there was Andre scooched down, with some seaweed draped over his eyes. He didn't realize that his whole rump was sticking up out of the water. Wasn't he surprised when Sue simply walked over and picked him up!

By the end of that first summer, Andre had become a waterfront character, especially to the fishermen, who aren't exactly fond of harbor seals. Seals eat fish. Fishermen, who catch fish for a living, think fish are only for people. But seals think fish are for seals. Still the Rockport men found it hard not to like friendly Andre.

They knew he enjoyed having his belly rubbed. When lobstermen rowed out to their boats, Andre would come up alongside and roll over, and the men would rub his belly with an oar.

He began taking naps in fishermen's rowboats or yachtsmen's dinghies, or he would sleep on lobstermen's wharfs. His favorite napping place was Howard Kimball's wharf. Howard Kimball, a lobsterman, sometimes found Andre a bit of a nuisance, for the seal had a habit of sleeping in Howard's rowboat when Howard needed it.

When Daddy learned that Andre was becoming a nuisance, he said to Howard, "Look, Andre needs a little discipline. If he's in your boat when you need it, just spank him with an oar."

"Now I can't do that," Howard said. "Andre's sort of a friend of mine."

"Then how is he going to learn what he can do or can't do. Andre's still just a kid. Don't you punish your kids when they're naughty?"

"Okay," Howard agreed, "maybe I'll baste him a little."

Andre's special friend was Howard's boy, Elray, who always took time from helping his father to play with our seal. All in all, it was a happy, lazy summer for Andre.

But summers don't last forever. As the days grew shorter, the nights became cold. The oak and maple leaves turned red and gold. Like us, Elray was off to school during the day, and there was no one for Andre to play with. Soon all the little boats were taken out of the water, and the floats hauled up on land for the winter. There were few places for Andre to sun himself and sleep.

Though Daddy didn't say anything, I knew he was worried. What would happen to Andre when winter came and the harbor froze over? Since he still refused to eat any frozen fish offered him, we knew he was catching fish for himself. But like the birds, fish go south in the winter, and soon there would be no fish for Andre to catch. By now, too, little Andre had become *big* Andre, much too big to be carried home to live in a bathtub.

In December, skim ice began to form around the edges of the harbor. That was no problem for Andre. He simply broke a hole in the thin ice, and as the ice got thicker he kept the hole open. Then he would pop up out of it to say hello to us as we stood on the shore and waved to him.

One night came a terrible storm, a wild northeaster. I lay in bed hearing the wind howl. Sleet and snow lashed against my windows like some wild animal wanting to get in. When I heard Daddy get up at dawn, I knew he was going down to the waterfront to look for Andre. I got up too, put on my warmest clothes, and went with him.

The seas were raging. The ice had broken up in the wind, and Andre's hole had disappeared. I could only see jagged ice floes grinding and buckling in the black water.

Daddy stood at the edge of the seawall staring out over the harbor, and finally turned away. "Andre's gone," he said. "Let's go home." Then he realized that "gone" wasn't the right word to use in breaking the sad news. "Andre's left," he said. That sounded better, but not much better.

6

Andre's First Journey

Maine winters are long and cold. The harbor freezes up tight. All through that February the snow fell. Plows piled snow so high along our road that I could see only the tops of the cars as they went by.

We had given up hope that Andre would return to Rockport Harbor. Still, on days when wind and tides broke up the harbor ice, we would go down to the shore and look out to sea. It was a long, sad winter for the Goodridge family.

One snowy day late in February, Daddy was opening his mail when he let out a yell and waved a newspaper clipping that had come in a letter. Mother took the letter and read it out loud.

It was from a newspaper in Marblehead, a little seacoast town near Boston. It told of a friendly little seal who had been loafing around the waterfront, and doing tricks for crowds gathered to watch him. The paper called this unusual seal, Josephine.

"Josephine!" Daddy snorted, "that's Andre! That's *got* to be Andre!"

He grabbed the phone and called the editor of the Marblehead newspaper, but the editor didn't seem very interested in a lost seal called Andre. Next my father phoned our cousin Jackie, who lives near Marblehead, but who had visited Rockport and knew Andre. Daddy asked him if he would take a close look at the friendly seal called "Josephine."

A few days later, Jackie called back to say Josephine had left Marblehead. "She — I mean *he* — hasn't been seen for over a week. But I'll bet dollars to doughnuts that was Andre," Jackie said. "What other seal would come up on a wharf and do tricks for people?"

"I guess we'll just have to wait," Daddy said, "and hope he's on the way home from his winter vacation."

So we waited. And waited. After March went by, the snow melted, and birds began flying in from the south. Then late in April, Daddy got a phone call from a man who had sighted a friendly seal off Newburyport, not too far from Maine waters. That was encouraging. If we didn't know where Andre was headed, we at least knew he was coming in the right direction.

And, waiting impatiently, we went down to the shore every day and looked hopefully out to sea. It was on the eighteenth day of May that another phone call came from a man in Rockland, a nearby town. "There's a seal asleep on McLoon's wharf. Could it be Andre?" My father jumped into his truck and took off.

We all gathered in the kitchen, along with the cat Martha and Toot, our lazy beagle. They didn't know what they were waiting for, but cats and dogs know when something is up. This was the hardest waiting of all. Was the seal in Rockland really our Andre?

When Daddy finally came back, he was lugging a great lobster crate, which he set on the kitchen floor. "It's Andre, all right. I recognized the rusty marking on his throat. But remember, he's been away a long time and may not even know us, or *want* to know us." Telling us all, including the cat and the dog, to stand back, he opened the crate and rolled Andre out on the floor.

The seal looked at us one by one. Then he humped over to the open cellar door and glanced down at his old quarters. Next he went bellywhopping around the house looking things over. Satisfied, he flopped down on the braided rug and fell asleep. For over an hour our travel-weary Andre slept. We spoke in whispers, and walked around on tiptoe, not wishing to disturb him.

Later the phone began ringing, friends and neighbors asking for news, newspapers calling for pictures and interviews. The whole village was delighted to have Andre safely home again.

When the seal awoke, Daddy carried him to the scales and weighed him. Sixty-four

pounds. Then he saw that one of Andre's flippers was badly chewed. Had he been in a fight? Had a shark attacked him? What worried Daddy most were the badly infected sores on his belly.

He immediately phoned a man he knew at the New York City Aquarium, and asked his advice. "Your seal may eat, and even gain some weight for a while," he said, "but the chances are he won't live."

Then my father phoned Mac, our animal doctor, who came right over. We all loved and trusted Mac. I certainly did. A busy doctor, he was never too rushed to tend to a wounded bird or animal. Once I took him my pet mouse and asked if he would put a splint on its broken leg. Some doctors might have said it was ridiculous to put a splint on a mouse's leg. But to Mac a little mouse was as important as a cat, a dog, or a horse.

"Hmmm," Mac said as he swabbed Andre's sores. Then he gave us a bottle of white tablets. "Give him nine of these a day until the sores heal."

I went to the door with Mac, not knowing what he meant by "hmmm." The word can mean something is very good or very bad. So I asked him straight out, "Is Andre really going to get better?"

"Well, your mouse got better, didn't he?" Mac said.

Andre didn't recover right away. We were almost ready to believe the aquarium man was right when the sores on Andre's belly began to heal, and in two weeks Andre was his frisky self again.

About this time, Daddy began to train Andre. Before the seal had come to live with us, Daddy had built a pound, a sort of floating playpen with a wire net bottom. The pound was intended for a dolphin, but my father never found a dolphin to serve as his swimming companion.

Since Daddy thought it would be better for Andre to live in saltwater while he was getting better, he put him in the pound moored just offshore. Several times a day he

would row out with raw fish, for now Andre was no longer free to fish for himself. I would go with him sometimes, and talk to Andre and play with him. It was like visiting a sick friend in a hospital in order to keep his spirits up.

When Daddy first realized that Andre liked to perform the little tricks he'd been taught, he would repeat them, mostly to please Andre. Soon he began to see that Andre was a natural show-off. One day at the pound, he said to Andre, "Maybe if I teach you a few more tricks, you won't go back to being such a tramp and a nuisance."

"Maybe," I said, "You can teach him to balance a ball on his nose like those seals in the circus."

"Those seals in the circus aren't harbor seals," he explained. "They are sea lions, and sea lions have been trained for hundreds of years. Nobody's ever tried to train a harbor seal."

"You mean they think harbor seals are stupid?"

"People know very little about them," my father explained. "But harbor seals enjoy a pretty easy life, you must admit. They don't have to work every day to make a living."

"Then if you want to train Andre, maybe you should make it seem — well, like a picnic."

Daddy looked at me and laughed. "I think you've got it, Toni. There are times when I think you are almost as smart as a harbor seal."

7

Andre Becomes A Celebrity

Andre had fun learning. Daddy made his training seem more like recess than school.

Often, when he went to the waterfront to feed the seal, I'd go with him. After teaching Andre a few simple tricks, like rolling over or clapping with his flippers, my father would give him fish as a reward. Then we would sit around and talk to him.

Andre acted as if he understood every word we said, as if he knew what was going on in my mind. I wasn't always sure what was going on in Andre's mind, though, and wondered how he would tell his own story if he could talk. Wouldn't it be interesting, I thought, if the big, bad wolf told the story of Little Red Riding Hood, or the three bears told the story of Goldilocks.

Everyone knows that animals don't talk like people; still, that doesn't mean they don't understand what is being said. Even Beulah, my stupid guinea pig, understood when I was pleased or displeased with her. But never had I known an animal like Andre who understood conversation.

As Daddy continued his training of the seal, more and more people began gathering at the seawall to watch. When they clapped, it was easy to see that Andre enjoyed the attention.

My father put a beach ball into the pen and taught the seal to swim about with the beach ball balanced on his nose. Then Andre would toss it into the air for Daddy to catch.

One day we heard a lady say, "Why that seal is almost human!"

Daddy said to Andre, "Perhaps we ought to tell the nice lady that you're only acting like a seal. Doesn't she know that seals were pushing things around with their noses a million years before humans learned to throw a rock."

So it wasn't at all hard to have Andre do things seals had been doing for a million years. Since seals leap into the air when they are feeling playful, Andre leaped through a motorcycle tire hung up in his pen. Seals are curious, too, so when objects were thrown into the water, Andre would chase after them and bring them back.

When a little boy in the audience yelled, "Hey, how do you catch a seal?" my father tied a knot on the end of a rope and tossed it into the water. Andre grabbed the knot and began tugging. Daddy hauled Andre onto the dock. "That's one way to catch a seal, son," he said. "But first you should get to know the seal."

I suppose a bear has to be taught to ride a bicycle, and a poodle has to be taught to stand on its hind legs and drink tea. Daddy didn't bother with that sort of nonsense. He asked Andre only to do things that seals have always done, except, of course, for a few silly things.

For instance, this is how he taught Andre to pretend to be ashamed of himself. One day, seeing Andre stretched out with one flipper over his eyes, he said, "Andre, aren't you ashamed of yourself?" — and right away gave him a fish. The very next time he

35

asked Andre, "Aren't you ashamed of yourself?" the seal stretched out and hid its eyes with its flipper.

It was just a wonderful game to Andre.

Knowing that a mother seal will slap the water with her flipper to warn her pup of danger, Daddy lighted a piece of paper and, when it was burning brightly, told Andre to put out the fire. Andre splashed water on the fire and put it out.

Perhaps you wonder how Andre came to pose for the camera. It happened this way.

Often after eating his supper, he would make a rocker of his body and nibble on his hind flipper. One day when he was doing this, Daddy said, "Pose for the camera, Andre," and the seal caught on right away. Once Andre learned a trick he seldom forgot it.

Of course, in the beginning he did get a bit confused now and then. Asked to clap, he might instead put his flipper over his eyes, but when not rewarded with a fish,he knew he'd got his signals mixed. He would then be put through all his tricks again until he did them perfectly.

So Daddy bought a whistle, and would blow it when Andre performed a trick right. Sometimes when Andre didn't hear the whistle, you could almost see his mind working: "What did I do wrong?" But he learned so quickly that my father now found he could try much more complicated tricks. Our seal was an eager pupil.

Perhaps you wonder how Daddy got Andre to say what he thought about Flipper, the dolphin. One evening when we were looking at the Walt Disney program about Flipper, and Andre was dozing on the rug beside us, my father suddenly turned to him and said, "Andre, what do you think of Flipper?"

Andre, knowing he was expected to do something, went through all his tricks. When Daddy didn't blow the whistle, Andre got a bit provoked. Now a harbor seal doesn't bark like a sea lion; instead he makes a snarling sound, blowing through his nose like a deer or a horse. He also snorts. Now Andre snorted in disgust, sounding very much like a bad-mannered boy making a raspberry.

"That's it, Andre!" Daddy yelled, and gave him a fish.

Of course Andre didn't get the joke, but he was certainly pleased with himself. Next day when Daddy asked him what he thought about Flipper in front of the crowd at the waterfront, Andre snorted. People clapped and cheered at his cleverness, and Andre all but took a bow.

As the seal grew smarter, my father found that he didn't need to use his whistle. A signal with his hands or a command in a quiet, everyday voice was enough. Even Daddy was astonished. To be astonished is one thing, but to be flabbergasted is something else. And one day in mid-summer Daddy was really flabbergasted by Andre's smartness.

That afternoon when we went to see Andre, he was resting, tired from leaping through the motorcycle tire. There he was draped over the rubber hoop like a limp Raggedy Ann doll. "You look ridiculous," Daddy said to him. "I'm glad Flipper isn't here to see you stuck in that hoop."

"Maybe," I said, "Andre's showing us what happened when he first started practicing."

"Hmmm," Daddy said. "That's an idea." Then he turned to the seal. "Maybe you'd like to show us how you jumped through the hoop when you were first learning."

Now this is the flabbergasting part. That night when the crowd gathered, my father got Andre to leap through the hoop a few times. Then he faced the crowd and said, "It took Andre hours of practice to leap through that hoop like Flipper, the dolphin. Now maybe he'll show you what happened when he was first learning." Then he walked over to the far end of the pound platform and said, "How about it, Andre?"

Andre went under. Then he took a running start and leaped into the air. He bellyflopped on the tire and hung there like a limp rag doll. The crowd roared, clapped, and whistled. Perhaps this was the day Andre decided what he was going to be when he grew up. He would be every bit as famous as Flipper, the dolphin.

Though he enjoyed all the cheering and clapping of the crowds, he loved first of all being a member of the Goodridge family. One summer day he showed us how much we all meant to him — and in the most flabbergasting way of all. I was on the dock, wearing my new blue coat with brass buttons, and reached down to pat him. He must have been

38

feeling impish that day, for he arched up, nipped one of the brass buttons from my coat, and tossed it into twenty feet of water.

I was angry and so was Daddy. Though he gave Andre a harsh scolding, no words could bring back my shiny brass button. But I was wrong. The next day when I went down to the waterfront, Andre was waiting there on the dock. He dropped the brass button at my feet.

8

The Long Hard Winter

Andre was now almost three years old, and growing up fast. I was growing up too, of course, but seals grow up more quickly than children. Soon he would be full-grown.

It often troubled Daddy that Andre, a harbor seal who belonged out in the open bay with other seals, preferred to be with people. Certainly Andre wouldn't know his seal relatives if he met them, and as surely the wild seals out on the ledges wouldn't know what to make of a seal who understood English and was driven about in an automobile.

With winter coming, we wondered what to do about Andre when the harbor froze over. This problem was solved when Luke Allen, who lived near the waterfront, offered his boat shed for Andre's winter quarters. The shed wasn't heated, but Andre now had a thick layer of fur and fat to keep him warm. But would he be lonely?

Daddy decided to build a tank and place it near a window so the seal would have a view of the harbor. To keep Andre happy, he would carry him up to our house to be with the family every few days. By the middle of November, Andre was installed in his winter quarters, and at least twice a week was brought home to stay overnight with us. Lifting Andre into the truck was getting harder every day for Daddy.

There is a story about a farm boy who was given a baby calf. "If I lift the calf every day as he's growing," the boy thought, "I'll get stronger as the calf gets bigger. And when the time comes I'll be able to lift a cow." I'm sure the farm boy never did lift a cow. Even lifting fat Andre caused my father to huff and puff.

During Andre's visits that early winter he became housebroken. It was not our idea; he decided to do so on his own. It came about this way. When Andre wet the floor, Daddy would go for the mop. The seal didn't like the mop slatting around him, and would snort. So my father slatted the mop even harder when Andre wet. The seal quickly realized that if he didn't wet, the mop wouldn't be whooshed around in front of his nose, so he stopped wetting.

Andre wasn't happy in his winter quarters. For me it was heartbreaking to go down to the boat shed and see him looking wistfully out to sea. By then the ice was forming, and it was too risky to set him free to face the Maine winter alone. Also he had lost his appetite. To lift his spirits Daddy would put him through a few old tricks, but even that failed to make him happy. Worse, he was becoming irritable. One day a state trooper, dressed in a crisp new uniform, asked if he could visit Andre. The seal took one look at the trooper, who leaned over to greet him, and slapped his flipper, soaking the police officer to the skin. The trooper gave Daddy a sour look and left, suspicious of my father's putting Andre up to the mischief.

And one day a nice lady brought Andre a present of four pounds of fresh smelt. "Is it all right if I feed him?" she asked.

Andre looked at the offered fish, then at the lady. His ungrateful thanks was a loud

juicy snort before he dived deep in the tank, disappearing from view. That night the Goodridges had the fresh smelt for supper.

As winter wore on, Andre grew more sad. One day a howling northeast storm struck in the afternoon, with high winds and blinding snow. Daddy knew we'd be snowbound for several days and wouldn't be able to feed Andre. So that evening he went to the boatshed and brought Andre home for the night, putting him in the cellar.

The next morning the whole world was white. It was still snowing, but most of the storm's fury had been spent. Daddy went into the cellar to feed the seal. Andre was gone. Then he felt a draft of cold air, and saw a small heap of snow on the cellar floor. Andre had apparently pushed out a pane of glass in the cellar window, and escaped into the white snowy world.

Daddy rushed upstairs. "Andre's gone!" he shouted.

I jumped out of bed. Like my father, I got into heavy boots and warm clothes. We had no trouble following Andre's escape route. The path in the deep snow led us down to the water's edge, where we stood looking out over the ice floes to the open sea. A foghorn was blatting away like a sick cow. There was no sign of Andre.

"Good luck, Andre," Daddy said softly, and slowly we walked back to the house where my father began phoning all the newspapers. He wanted the word out that a friendly seal, who didn't know he was a seal, had escaped and was on the loose.

It was well he did get the word out. A few days later the harbor master at Round Pond, a coastal village fifty miles up the coast from Rockport, telephoned to say that a friendly seal, who might be Andre, was entertaining a group of delighted children there.

Daddy rushed off in his truck, but by the time he got to Round Pond, Andre had departed.

The harbor master told my father a frightening story. "Your seal had a close call, you know. I was in a store when a lobsterman saw the seal on the shore, and went for his rifle. He had the mistaken notion that seals eat lobsters. "Look," I said to this fellow, "doesn't it strike you as a mite odd to see a seal playing with children? I'll bet he's that Rockport seal called Andre. Now you just put that gun away."

Daddy thanked the harbor master for saving Andre's life, and asked him to call if the seal showed up again. Then he went home to wait.

Andre did show up in Round Pond the next day to give another performance, but this time he waited for Daddy, who picked him up and brought him home. During the drive back to Rockport, Andre fell peacefully asleep in the back seat of the car, not realizing how close he had come to death.

9

Andre's Friend Elray

On one of the first fine days of spring, Andre's pound was put back into the harbor, a festive occasion in Rockport. Sea gulls screamed and swooped about; dogs romped. Lobstermen were scraping and painting their boats, getting them ready for another fishing season.

Andre was joyous. Around and around the little harbor he went, diving and leaping. Everyone along the waterfront watched as he frolicked like a colt in spring pasture. Willingly, he shot into his pound, happy to be in his floating home again. He loved and lived in two worlds, the world of people and the wild world of the sea. It was the beginning of a wonderful summer.

Best of all he liked Daddy's company in the water. When my father went scuba diving and released Andre from the pound, he didn't have to call to him, as a man calls to his dog, for the seal was always close by. Underwater was Andre's world. Sometimes it seemed that now he thought it was Daddy who needed protection.

When another fall rolled around, Daddy decided to put Andre back in the storage shed once again. He disliked doing it, knowing how the seal hated being cooped up. In the tank for less than a month, Andre refused to play, nor would he eat. By early December it became clear that Andre would languish and die if left in the storage shed. Free in the open sea Andre would face certain dangers, such as sharks, or he might be shot by lobstermen who didn't appreciate fish-eating seals. But Andre would have to take his chances.

Daddy called the newspapers, wanting people up and down the coast to know that Andre would soon be set free. Newspapermen came with their cameras and took pictures as my father lifted Andre out of the tank and set him on the floor. "You're on your own, Andre," he said. "Get along now."

Andre didn't hesitate. He went humping out of the shed and made a beeline for the water, where he bellywhopped into the sea. He went porpoising around the harbor in great joyful leaps.

He turned back only once. I was there on the shore with my parents. He came towards us just under the water and, facing us, made one last leap and spanked the water. Then with a final slap of his hind flippers, he headed for open water and was gone.

We couldn't imagine what would become of him. Freed from the dark shed, would he enjoy a few days of hobo life and then come home? Would he go south to Marblehead and visit his friends there? Perhaps he was gone forever.

A day later, Daddy received a phone call from Rockland, where Andre had been seen playing in the harbor. My father drove to Rockland and found Andre in the shallows. Though the harbor was freezing over by then, the seal had found a little cove that was free of ice. When Daddy spoke to Andre, the seal replied with a snort. My father understood the snort all too well. It meant, "Don't bother me. I'm not ready to come home yet."

We had only one more report of Andre, this time from a man in the harbor town of Friendship, further along the coast. Andre had stopped off for a short visit there and had left.

At home, we waited and worried. Weeks later, I said, "Maybe we should call the Coast Guard. Don't people call the Coast Guard when someone is missing at sea?"

"I'd feel silly asking the Coast Guard commander if he'd seen any friendly seals lately," Daddy replied, but all the same he called the Coast Guard. The commander didn't think the request at all silly. "I'll ask all the Coast Guard stations to be on the lookout," he promised.

Then one day Luke Allen phoned. "Hey, Harry," he said. "Andre's back in the harbor!"

My father hopped into his truck and rushed down to the waterfront. Down on the shore rocks, Andre, recognizing the sound of Daddy's truck, was craning his neck, looking this way and that. Daddy was appalled at the seal's appearance. Andre was pitiably thin, and his coat had lost its healthy shine.

My father called out, "Stand by, boy!" and he rushed home to get some fish.

Happy to see Daddy, Andre was even happier to see some fish. Had poor fishing in the sea, or loneliness, brought him home to us? Perhaps both. Now we were quite sure he wouldn't wander far again.

Over the summer, Andre played in the harbor, caught his own food, and he swam along when my father went scuba-diving or diving for scallops. But what seemed like a perfect arrangement for all of us turned out to be not quite perfect.

No longer a cute, cuddly pup, Andre, at four years old, weighed 150 pounds. He still loved having his belly scratched with an oar, and when fishermen going out to their boats didn't give him a friendly scratch, he took to nipping at their oars. One morning, a lobsterman named Dave started out in his dory to tend his traps. Andre came alongside and began to plague him. Dave, pushing heavy oars against the wind and the tide, found it a nuisance to have a playful seal grabbing at them. Finally, Dave gave up and rowed back to the dock. Seeing Daddy, he said irritably, "Look, Harry, you've got to do something about that pesky seal!"

Right away, Daddy got into Dave's dory and began rowing out into the harbor. Every time Andre grabbed an oar, my father thumped him with it. Sometimes he had to thump him half a dozen times before the seal understood that this was not a game.

There was another problem, too. Andre rested or slept when and where he felt like it. No one had minded when the seal was a small pup, but now his great weight on the gunwhales, as he crawled in for his naps, would often so tip the boat that it would fill with water. To Andre, apparently, rowboats were meant for seals to snooze in. Certainly, visiting yachtsmen didn't find anything amusing about a seal using their dinghies for a bathtub.

In turn, Andre didn't find it amusing to be disturbed from his napping; he'd simply remain stubbornly in the boat and snort loudly. Time after time that summer, Daddy was called to the waterfront to remove Andre from someone's rowboat.

Daddy didn't quite know how to handle this problem. Of course, when fall came, the yachtsmen would be gone, and the floats hauled up on shore. Andre might even

learn to bask on shore rocks as seals are supposed to do. But Andre didn't like shore rocks. Instead, he set up winter quarters on Howard Kimball's wharf.

Howard was the only man in the harbor to fish year round. He had a skiff, and a wharf crowded with all sorts of things to sleep on — ropes, lobster pots, chains. And best of all for Andre was Howard's boy, Elray, whom he loved. It didn't take long for the seal to learn what time Elray returned from school. Each afternoon he would be there on the Kimball's wharf waiting for his friend.

It was a good thing that Elray loved Andre, too, for the seal could be a pest and a rascal. For instance, Elray's father caught crabs which he stored in a floating crate just offshore, and one of Elray's daily chores was to get the crabs out of the crate and take them to the house for his mother to prepare for market.

When Elray would take one handle of the crate and try to haul it ashore, Andre would get his teeth in the other handle and pull the other way. Then if Elray was too busy with his chores to play, Andre would get annoyed and splash the boy with his flippers, or nip at his pant legs. Used to summer crowds watching him perform, Andre expected human attention. When he didn't get it, he demanded it.

So Elray would take time from his chores, throw sticks in the water for the seal to chase, or hold one high in the air and have Andre jump for it.

It was a great day for both of them when Elray learned to scuba-dive. After all, while a seal likes the land as a place to rest and snooze, the sea is its natural playground. When Elray would get into his rubber suit and dive, Andre would shoot between his legs. Together they learned to play leapfrog underwater, and together they would dive deep into the harbor looking for scallops.

When it came time for Elray to go to the house for supper, Andre would follow him up on the rocks. "See you tomorrow, Andre," Elray would say, waving goodby.

Andre then lay on the rocks looking up at Elray's house, hoping the boy would come out again to play. When night fell and the lights went off in the Kimball house, Andre would slip back into the dark water to await another day.

10

How Smart Is A Seal?

It was bound to happen sooner or later. Though Andre had many friends, a few fussy people doubted that the harbor was the proper place for a free-living, full-grown seal. One of them was a man rowing his little boy in the harbor on a fine spring day. Suddenly, Andre appeared from nowhere and, leaping up out of the water, gave the boy a big wet kiss on the cheek.

Frightened, the little boy burst into tears. His father went straight to the fishery warden, and he complained that his little son had been attacked by a seal. A few days later the warden spoke to Daddy. "Harry," he said, "we're going to have to do something about Andre."

There was only one thing my father could do. He would have to keep the seal in his floating pen until the fall when all the boats were hauled up for the winter and there were no little boys around to be kissed.

Daddy explained the situation to Andre as best he could, and ended up by saying, "Sorry, old fellow, but that's the way it is."

By then Andre was old enough to understand that he was a seal, and that people were people. People and seals look at things differently. He had learned to accept this philosophically. At least, with his gray whiskers and bald head, he had a way of *looking* philosophical.

But if Andre was going to be confined to his pen all summer, his pen would have to be bigger before another summer rolled around. Of course we still didn't know how Andre was going to spend his winters. Daddy could only let him go out to sea and hope for the best.

He waited until November. On a cold blustery day he rowed out to the pen and opened the gate. "Okay," he said to Andre, "school is out. Have a nice time."

"Be sure to drop us a postcard," I said, trying hard to be funny.

Andre didn't hesitate. He slipped out into the open water. After several quick turns around the harbor, off he went, but not for long. A few days later, he swam back into the harbor looking for us. We could see he expected to be fed.

Was Andre getting lazy? He was quite capable of finding his own fish, but, of course, it might be that fish were scarce. While wild harbor seals don't go south with the birds in the winter, they sometimes have to wander a distance looking for food. Maybe Andre thought to himself, "I know where I can get some fish. Those fish *he* feeds me aren't exactly fresh, but they are better than none at all."

That winter even fishermen were having trouble catching fish in the bay. One of them used to call my father when he had some to spare for Andre, but during the cold winter he didn't. Often Daddy had to drive ninety miles to Portland to find fish to put in the freezer for Andre. Then, every evening, he had to go down to the waterfront in all kinds of weather with a bucket of thawed fish. Andre's feeding time was seven, and don't think that sly rascal didn't know when it was seven o'clock!

In early winter, it's pitch-black at that time of night. When after a cold snap the harbor would freeze over, we were never sure where Andre would find an opening in the ice. One bitter night in January he was nowhere to be seen. It was spooky and black as a cellar hole down there on the waterfront. The ice scraping against the rocks made creepy, animal noises. We called and called. No Andre.

Then Daddy had a brilliant idea. "Hey, Andre," he called into the darkness, "what do you think of Flipper?" Out of the darkness came a loud and juicy snort. So Andre got his supper that night — and we got home before we froze to death.

Some nights were so dark that we couldn't see Andre even when we'd located his hole in the ice. When Daddy would push a fish into the icy water, the unseen Andre would take it from his fingers. When all the fish were gone, my father would say,

54

"That's all, old fellow." He would wait a moment, then feel into the hole with his hand. Andre would be gone out to sea under the ice.

The seal never missed his feeding time that winter. Some nights the harbor was frozen all the way out to the open bay. How did he manage to swim all that distance under the ice without air? A seal must have air to breathe, just like people. Andre could rest at the bottom for as long as fifteen minutes without coming to the surface to breathe. But when a seal is swimming, he uses up oxygen much faster than when he's resting.

It was a mystery. It's a mystery, too, how birds find their way back to their nesting grounds in the spring. And why is it that sea gulls can stand for hours on ice floes without getting their feet frozen to the ice. How does a spider make the exact same web time after time without a pattern. Mysteries are things we just don't understand.

There were many other puzzling things about Andre. He loved scuba divers. When one of them would go into the water a half-mile away from where Andre was swimming, he'd find the diver in a flash. And he'd want to play. Maybe a man in a black rubbery suit looked and acted like another seal to Andre.

One afternoon, Daddy and some friends went in a boat to dive for scallops off Ram Island. Ram Island is four miles' distance from Howard Kimball's wharf, where Andre was sleeping that day. The men were in the water for no more than ten minutes when a familiar whiskered head bobbed up in their midst.

There was Andre. He'd come four miles to the spot, straight and swift as a bullet!

When people use their brains to figure things out, it is called intelligence. People like to think of themselves as so much smarter than animals. But one day Andre showed us that seals can figure things out, too.

Daddy had run out of his supply of frozen herring, and decided that he'd have to feed the seal some mackerel. Dumping a pailful on the deck of Andre's pen, he received

56

for thanks only a snort of disgust. "Sorry, old fellow," Daddy said, "but mackerel is all we have today."

Andre took one of the mackerel in his teeth and, ripping it open, slapped it about a bit. Then he pushed it uneaten into the water of the pen, and watched it sink to the bottom of the wire cage. Suddenly his back arched. He slid into the water and came up with a live, flapping harbor pollock.

We didn't know what was going on at first. And when we realized what Andre was up to, we could hardly believe it. For next Andre mashed up yet another mackerel and pushed it into the water. In a moment, he went into a dive and came up with a second flapping pollock.

The mashed mackerel, it seems, was luring those little harbor fish in through the wire mesh of his cage. Andre was using fish he didn't like to catch fish he *did* like!

Astonished, I exclaimed, "He's fishing with bait!"

Daddy was too dumbfounded to say anything.

And I could just imagine Andre thinking disdainfully, "So I use bait to catch fish? What's so amazing about that? Do you think I'm stupid or something?"

11

Trudy

Andre, going on five, was much too heavy to be carried to the house regularly. Though we all visited him often, there were long hours in the summer when he was alone in his pen.

He needed company. He loved people, thoroughly enjoying the applause of crowds that gathered to watch him from the sea wall on summer evenings. Still, he was a harbor seal and must have yearned for seal company.

Surely he had come upon brother seals on his winter wanderings, but had they accepted him? Did they think him strange and different, as any seal brought up with people would be?

We guessed as much, for sometimes Andre would come home from his winter jaunts with torn flippers and nasty wounds. While sharks are a seal's only enemies in the open sea, by now Andre was too big to tempt a shark. Clearly, Andre had tried to make friends with wild seals and had been driven off.

If he couldn't find seal company in the ocean, the next best thing would be to bring seal company to him. Best of all, a girl seal. And that was what my father had in mind when he went to Boothbay Harbor early the following summer. He knew a man who kept some young seals in a tank at the State Marine Laboratory there.

Daddy explained his problem. "What I'd like," he said, "is a rather dumb female seal."

"There's no such thing as a dumb seal," was the answer.

"Well, then I'd like a young female seal who isn't *too* bright. Now that I've trained a smart seal, I want to see what I can do with a seal who isn't so smart."

"Hmmm," the man said thoughtfully as he looked over the young seals in the tank. "That cute little one over there is the easiest to catch. She'll swim right into the net. So she's *got* to be the dumbest."

And that was how Trudy came to live with Andre. She was a great success, and Andre was delighted with her. When he tried to nuzzle and cuddle her, Trudy would nip Andre playfully. Then the two would chase one another round and round the pen like kids in a schoolyard.

Andre would tease little Trudy as big brothers tease little sisters. Sometimes he'd manage to steal Trudy's fish at feeding time, but quite soon Trudy was giving as good as she took. Faster in the water than Andre, she could beat Andre to the fish. In fact, very soon Daddy had to step in and make sure Andre got his fair share of food.

Before a month was out, it was Trudy who was doing the teasing. Daddy had taught Andre to push a beachball with his nose and toss it into a hoop tacked up at one end of

the pen. Trudy would wait until Andre was ready to shoot a basket; then she would whack it away with her flipper. Andre would try again, only to have his mischievous girl friend flip the ball away at the last instant.

But you can be sure Andre wasn't outwitted for long. He would pretend to ignore the beachball as it floated in the water. He'd wait until Trudy went for a fish that Daddy tossed into the pen. Then the instant Trudy's back was turned, he would make a dash for the ball and shoot the basket.

We all hoped that the great friendship would go on forever. Sadly, it didn't turn out that way. Daddy released Trudy and Andre together when the summer was over, hoping the two friends would stay together and return together. Free in the harbor, however, Trudy looked back just once. She turned to the open sea and the sea closed over her. Andre returned alone a few days later, and we never saw little Trudy again.

Now and again that next year, Daddy would come upon a wild seal in the open bay,

who seemed to show a special interest in him. Always, he turned his boat away. If Trudy had returned to the open sea and her own kind, it was because that was the way she wanted it. It was best, my father said, that things be left that way.

12

The Honorary Harbor Master

Andre had been an adorable baby, then a lively and mischievous youngster. Now he was a dignified gentleman seal. Though he still enjoyed the applause of the crowd, he had become his own boss. And there were times in the summer when he just didn't feel like clowning for an audience.

One day, a relative of Daddy's phoned to say he was bringing a group of friends to the Rockport waterfront. Having seen Andre perform, he said, "They don't believe a word of what I've told them about Andre. I'm counting on that seal to do his whole bag of tricks."

That afternoon Andre looked up at the skeptical faces on the sea wall. He decided he didn't have to prove anything to anybody. He sat there like a bump on a log and refused even to clap his flippers. My father was embarrassed.

When Andre was a pup, Daddy had looked out for him. Now when my father went diving, Andre looked out for him. After all, seals had lived in the sea for millions of years, but people were just learning to be at home in the water.

One spring afternoon, Daddy was diving in the harbor, looking for a lost mooring. Deeper and deeper he went into the dark, murky water, much deeper than it was wise to go. Suddenly, something grabbed him from the rear. Startled, he reached for the knife he always carried when diving, and swung around, not knowing what to expect. There was Andre in a high state of agitation. He began pushing Daddy toward the surface, telling him that it was all very well for a seal to dive that deep, but for a human being it was dangerous.

I remember the day Andre refused to wear a harness. For a time my father thought that seals should be trained to perform useful tasks. After all, St. Bernard dogs have been trained to pull a children's wagon; for centuries elephants have been trained to move great logs with their trunks. Daddy had trained a number of seals to wear collar harnesses and carry lines out to boats. In most matters Andre was very obliging, but he drew the line at wearing a collar.

He looked at the thing in disgust, flatly refusing to have one put over his neck. And no amount of coaxing or stern words would change his mind about that.

My father impatiently threw the collar on the deck and went ashore, but immediately regretted his childish behavior. Why should he insist upon Andre's doing something beneath his dignity? Andre was a friend, not a servant. When Daddy returned to the pen to offer his apologies, Andre was waiting in portly majesty, with the collar around his neck. He had put it on himself, apparently just to please my father. Daddy never again asked him to wear a collar.

Andre's fame had been growing with each passing year. The summer crowds at the sea wall grew larger and larger. School children came by bus from all over Maine to see

him; boatloads of youngsters came from outlying islands to watch his performances. Not only did Andre remember all the tricks he was ever taught, he thought up new acts to please the crowds. There was the sea gull act, for instance.

Sea gulls always arrived at feeding time, hoping for a free meal. Daddy would toss a herring out over the water. Andre would plunge off, snapping up the fish before the gull could reach it. If he failed in this, he would simply leap out of the water and snatch the herring from the gull's beak. He seemed to enjoy proving that a harbor seal was smarter and faster than any sea gull that ever lived.

One day during a performance, a little rowboat got loose and began drifting away from the dock. Daddy yelled, "Andre, get out there and bring the boat back."

Like a porpoise, Andre leaped forward, took the line in his teeth, and brought it back to the dock. Afterwards, this became one of his favorite tricks. Since he swam underwater when he was towing the boat in, it was quite a spooky sight to see a little boat moving along without a motor or a man at the oars.

Finally the great day came when Andre was named honorary harbor master of Rockport. For some years my father had been Rockport's official harbor master. Now wasn't it fitting and proper, our town manager thought, to make Andre his honorary

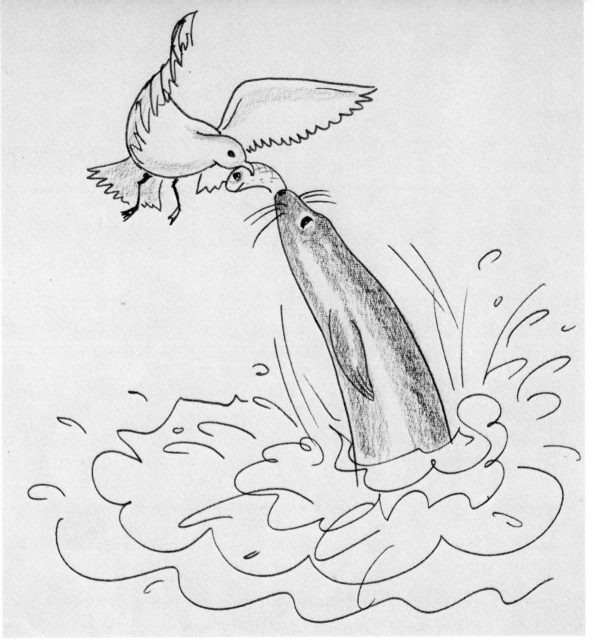

assistant? Andre accepted his badge with quiet dignity. He liked the honorary part, for it meant that he wouldn't be called upon to do any hard work. That suited him just fine.

Never before had a harbor seal been so honored. But by no means was Andre ready to rest upon his laurels. It was time now for his Great Adventure.

13

The Old Mariner

Of all the seasons, summers are remembered best, and there were many happy summers before Andre set forth on his great sea adventure. There were some bitter winters as well. And it was during those long winters that Andre learned about the great watery world that lay beyond the bay where he was born.

He was off on his own for weeks at a time. Like the voyages of Maine sailing-ship captains, his explorations took him to far places. Like them, too, he learned to know the wind, the tides, and the currents. Old-time sea captains sailed by the stars, but no one knows how a seal finds his way home. Andre had a seal's way of knowing where he was and where he wanted to go. Certainly he knew his way home, and home he came with the spring.

Now and again, he would return to see us deep in the winter when the wind and tide broke up the ice in the harbor. One cold day in the middle of February, he came home in a style that befitted an honorary harbor master.

Howard Kimball, whose lobster boat had been icebound for a week, was working on his wharf that afternoon. Earlier, he had called the Coast Guard, asking them to send in an icebreaking vessel to clear a channel in the harbor that would allow him to fish for crabs.

As the boat approached, he waved to the captain and thanked him, and the Coast Guard captain waved back as he went by. Then he shouted something that Howard couldn't hear above the sound of crunching ice, and jerked his thumb towards the stern of the boat.

"Gorry," Howard told Daddy later, "I couldn't believe my eyes! There was Andre in the wake of the vessel. He was swimming in the open channel and putting on quite a show, leaping and diving among the ice cakes. You might have thought the icebreaker had been brought in to clear a channel for *him.*"

His winter visits were brief. After saying hello, he would take off again for parts unknown, but each spring he arrived when the ice had melted. How he enjoyed romping around as the boats were put over and the floats set out. Following a long winter alone in the open sea, he seemed to hunger for human activity and people's company.

Then, just as everything seemed to be working out perfectly, Andre got himself into serious trouble. It happened in the spring of his fourteenth year, a few weeks before his pen was put in the water for the summer. Free in the harbor, he came upon two young men in a canoe, and decided to play. In his exuberance, he capsized the little boat and the two men were forced to swim ashore. Luckily, they were good swimmers, but what if they had been unable to swim?

A few days later, a man and his wife were in a kayak out in the bay. Now a kayak is even more easily tipped over than a canoe. Andre, nipping at the paddles and diving under the delicate little boat, so frightened the man and his wife that they didn't think they would get home alive.

Scuba divers became another problem. Scuba-diving was now a very popular sport, and more and more divers were in the waters around Rockport, many of them strangers. Imagine being deep down in the water and suddenly finding yourself hugged by a two-hundred-pound seal! Daddy was fearful of what might happen should a diver panic. He was also worried about Andre. Many of the divers carried spear guns. And surely these underwater hunters were not used to seals that wanted to play leapfrog. They might think that they were being attacked.

Daddy got the pen over quickly and placed Andre in it. My father knew he must now seriously think about Andre's future. According to the law, Andre was my father's responsibility. All of us Goodridge children were growing up and one day we would be off on our own. But it was clear by then that Andre wasn't going off on his own for good, and as long as he lived he would be my father's responsibility.

One of the first questions people ask Daddy is how long does a seal live? He tells people that no one knows how long a seal lives in the wild; in aquariums, seals have been known to live to the ripe old age of forty. Even if our Andre lived to be only thirty, that was a long time for someone to be responsible for his actions.

Having thought about this all that summer, my father reached a decision in late August. He phoned Dr. Garibaldi at the New England Aquarium in Boston. "I'm Harry Goodridge in Rockport, Maine," he said, going right to the point. "You know, the fellow with the seal. How would you like to have a winter boarder?"

Dr. Garibaldi certainly knew of our famous seal. "If you're talking about Andre, we'd love to have him. We'll come up with a truck and get him whenever you say the word. And, of course, we'll truck him back in the spring."

Then my father said something that startled us all. "You won't have to deliver him in the spring. I want Andre to swim home to Rockport."

I'm sure Dr. Garibaldi was as startled as we all were. "Are you serious?" I said to Daddy, when he hung up the phone.

"Of course I'm serious," he said. "If Andre is going to be penned up both in Rockport and Boston, don't you think he needs some freedom? He's smart. If he wants to come home, he'll find his way."

"But, but," I stammered, "what if he doesn't want to come home?"

"We'll leave that decision to Andre," was the answer.

14

The Great Adventure

On a raw, gloomy day in early November, a small crowd had gathered at the waterfront by noon when the truck arrived from the New England Aquarium to take Andre away. Out of the van stepped Bob Anderson and a pretty young lady named Annie Potts, members of the aquarium staff who would watch out for Andre in Boston.

Everything went smoothly. The men took a slatted carrying cage in a rowboat out to Andre's pen. Andre plopped into it without any fuss. He seemed relaxed and unconcerned as he was carried up the steep ramp to the waiting truck.

When he was loaded aboard the truck, my father said, "You're going on a vacation, Andre," and he put his head against the slats of the crate so Andre could say goodbye. Andre sniffed Daddy's hair, for that is the way seals say hello to friends. Now this was hello and goodbye.

The rear doors of the van clanked shut, and the truck went off up the steep hill to the village. Andre was gone.

Everything was going to be all right, I told myself. Andre would be safe. He would have other seals to play with, and children from all over the world would come to see him. At least I hoped so. All the same, it was a relief to learn a few days later that Andre had arrived safely in Boston.

"Everyone has fallen in love with Andre," the aquarium man said on the phone. "In fact, he has taken over the place. There's only one small problem. Another seal here, called Hoover, is sulking, because Andre likes his lady friend."

Later that winter, Daddy flew to Boston to see for himself how Andre was faring in the big city. We were all waiting eagerly for news when he stepped into the house that evening. "He seems to be having the time of his life," he reported, "swimming around in the tank with his lady friend. When I called his name, his head popped up and over he came to see me. Satisfied that I was the one and only Harry Goodridge, he went back to join his friend."

"But doesn't he miss us?" I asked.

"Oh, I'm sure he does," my father said. "We'll know in the spring just how much, won't we?"

When spring came at last, Andre was to be released to make the long swim home alone. He would be free to roam the seas, or to return to the only family he had ever known. I didn't dare guess what he would decide to do. Now it was entirely up to him.

The date for his release was April 26. The aquarium people thought it would be best to set Andre free at Marblehead Harbor, twenty miles north of Boston, where he would meet less ship traffic. Daddy agreed readily, for Andre had once visited Marblehead when he was a seal pup. Surely he would know his way home from there.

My father notified all the newspapers about the coming event, wanting to be doubly sure that everyone along the coast knew that a friendly seal called Andre would be swimming north to Maine.

Of course, some people doubted that our well-traveled seal would make it home to Rockport. I was sure he would, for I kept remembering the day he dived for my coat button and dropped it at my feet. Whatever else he was, he was smart.

Daddy flew to Boston on that special day and rode with Andre in the van that took him to Marblehead. A forty-foot fishing boat was waiting at the dock, as were reporters and photographers. After Andre was carried aboard the boat in his cage, the fishing-boat captain freed his lines. Everyone was excited, but not Andre. He had lived such a strange life that he didn't know it was unusual.

When the fishing boat had sliced a few hundred yards out into the harbor, the aquarium man said to my father, "You say where and when to set him free."

"What's wrong with right here?" Daddy asked, and opened Andre's cage. Andre humped out, looked at Daddy, and *swoosh* into the sea he went. Just once he looked back. "Okay, Andre, go on home," my father ordered.

Andre dived, flicked his hind flipper and was gone.

When my father arrived home late that night, he hung up his hat and lighted his pipe. "Now he's on his way," he said. Nothing more. Only when I was ready for bed did I ask the question that was most on my mind. "How long will it take a seal to swim home?" I thought it a very good question, but I didn't get a very good answer.

"How long did it take you to walk home from school when you were a little girl?"

"Well," I said, "I *could* walk home in twenty minutes. But some days I stopped for an ice cream cone. It all depended."

"With Andre, too," Daddy said, "it all depends."

So once more the waiting began. It seems that my story of growing up with Andre is filled with waiting. But this time the waiting was brief. The first report came the very next day. A man phoned from Kittery Point, fifty miles north of Marblehead. "You that fella with the seal?" he wanted to know.

"Yes," Daddy said, "I'm Harry Goodridge."

"Think I saw your seal," the man said. "He was resting on my dock. I patted him."

"If you got close enough to pat him, that's Andre all right!" my father said.

There was no word for the next two days. On Tuesday morning the phone rang. A girl who lived just around the point from Rockport had sighted a seal. Daddy thanked her though he doubted she had seen Andre. How could a seal possibly travel all that distance in four days!

He was wrong. An hour later a man called saying he had sighted a seal at Owls Head, nearer home. Then another man called saying he'd seen a seal in Rockland Harbor closer still to home.

Daddy dashed for the freezer. Filling a pail with fish, he rushed down to the waterfront where he waited in vain for an hour. He was home having lunch when a neighbor telephoned. "Harry," she exclaimed, "Andre's home! He's asleep in our rowboat."

We never did finish lunch that day. In a few minutes we were in my father's boat and chugging across the harbor. Sure enough, there was weary Andre curled up in a

rowboat. At the sound of our outboard motor, he raised his head over the side and looked at us owlishly.

"Hello there, old fellow," Daddy said, offering him a fish. "How was the trip?"

Andre showed no interest in being fed. For four days he had traveled on a homeward course and now only wanted to sleep. Later that day, after a long rest, he slid gratefully into his waiting pen, even as people all over the country were reading of him in their newspapers. ANDRE MAKES IT HOME!

We sat and talked to Andre until the sun went down and the sea gulls were flying out to their night roosts on the islands. Then I knew for sure that this was home to Andre, and I'd never need worry about him again.

15

There Will Always Be An Andre

Alice, of *Alice in Wonderland,* had this to say about telling a story. "You start at the beginning and when you get to the end you stop," she said.

This isn't the end of Andre's story, of course. Now, however, Andre is famous. Everyone in the country knows about the harbor seal who swims home from Boston each spring to spend the summer in Rockport, Maine. As far off as California, people read in their newspapers that Andre has left the aquarium in Boston for the long swim home. And thousands of people await anxiously until they learn that he has again arrived safely in our little harbor.

We're never quite sure about Andre's exact time of arrival. One spring, he made a record trip of three days; some years he decides to dawdle along the way, sight-seeing and making side trips along the coast. One spring, it was a good two weeks before he showed up in the harbor.

But we no longer worry about him. Since everybody knows about him, people who see him along the way telephone us. Always he is glad to be back, and always he slips happily into his pen.

Andre will soon be twenty years old. His coat is turning gray. He is wise in the ways of the sea, wise about people. When he lies dozing in the sun, does he think wise thoughts, too? As I said at the very beginning of this story, I am never quite sure what goes on in Andre's head.

Does he know that he's as famous as Flipper, the dolphin? As celebrated as Smokey, the bear? Does he know that he's the only harbor seal in the whole wide world who lives in two worlds and swims home to Rockport from Marblehead each spring?

He should have guessed he was something quite special one glorious day in October. A famous sculptor carved a two-ton granite statue of him and donated it to the town of Rockport. The real Andre was there the day of the unveiling. Bands played, children cheered. And there, carved in stone larger than life, was Andre looking seaward towards the island where he was born.

I don't suppose it matters that Andre didn't really understand what the fuss was all about. He dozed. What matters is that a beautiful statue of him will be there looking out to sea forever.

Now, come what may, there will always be an Andre.